At The Park

by PJ Dietz

I found the slide.

I found the **swing**.

I found the **bench**.

I found the table.

I found the **tree**.

I found the flowers.

I found the bridge.

I found the steps.

I found the fence.

I found the fountain.

YY: Skills Card

Reader ______________ Room ________

Strategies	Date:	
Listen to the title and first page.		
Use the title and first page to figure out the rest of the pages.		
Use my finger to point to each word as I read.		
Use the picture and the first letter to figure out the words.		
Use the spaces to separate words.		
Tell someone what the book was about.		
Read at home every night.		

I can get my mouth ready for

b	c	d	f
g	h	j	k
l	m	n	p
r	s	t	z

I can read these words

I	love	a	cat
is	The	the	A
stop	Stop	go	Cat

Can you match the words to the pictures?

slide

flowers

swing

fountain

bench